Cracking Pediatric Neurology Vignettes
First Edition

Paul Edward Kaloostian MD,   Sean William Kaloostian MD, Carolyn Louisa Kaloostian MD, MPH

Printed in the United States of America
First Printing, 2012
ISBN: 978-1-300-17550-6

ISBN 978-1-300-17550-6
9 781300 175506
90000

About the Authors:

Paul Kaloostian MD: Matriculated through Thomas Haider Accelerated Biomedical Sciences B.S./B.A./M.D. Program with Undergraduate work at University of California Riverside and Medical School at David Geffen School of Medicine at UCLA. Currently, he is Chief Resident at University of New Mexico Medical Center in the Department of Neurosurgery. He will be attending Johns Hopkins Medical Center for his Complex Spine and Spinal Oncology fellowship in 2012. He has a passion for treating the underserved and is fascinated by the immense cultural diversity in New Mexico. He is fluent in three languages: English, Spanish, and Armenian. He is an avid pianist and concert clarinetist and composes classical and Armenian Folk music.

Carolyn Kaloostian MD/MPH: Matriculated through Thomas Haider Accelerated Biomedical Sciences B.S./B.A./M.D. Program with Undergraduate work at University of California Riverside and Medical School at David Geffen School of Medicine at UCLA. Currently, she is chief resident in the Department of Family Medicine at University of Southern California Medical Center. She is currently obtaining her MPH degree at UCLA during her residency. She will attend UCLA Medical Center for her geriatrics fellowship in 2012. She has a passion for treating the underserved populations. She is an avid ballerina having performed in multiple national performances. She is fluent in three languages: English, Spanish, and Armenian.

Sean Kaloostian MD: Matriculated through Thomas Haider Accelerated Biomedical Sciences B.S./B.A./M.D. Program with Undergraduate work at University of California Riverside and Medical School at David Geffen School of Medicine at UCLA. He is currently a resident at University of California at Irvine specializing in Neurosurgery. Of note, he is a National Rhodes Scholar Finalist and Varsity Baseball player. He has also completed many marathons with competitive times. He has a passion for treating the underserved communities. He is fluent in three languages: English, Spanish, and Armenian.

Editors-in-Chief

Paul Kaloostian MD- Fellow and Instructor, Department of Neurosurgery, Complex Spine and Spinal Oncology, Johns Hopkins Medical Center
Carolyn Kaloostian MD/MPH-Fellow, Department of Geriatrics, UCLA Medical Center
Sean Kaloostian MD-Resident, University of California at Irvine Department of Neurosurgery
William Kaloostian MD-Hospitalist, Internal Medicine Clinical Professor, Director of Short Stay Observation Unit, Kaiser Permanente Medical Center, Los Angeles, Ca.

Special Thanks/Dedications

This was a Herculean Task that could not have been possible without the efforts of many people. Special thanks to Eddie and Nanny for instilling with us the desire to teach and promote education, as well as a duty to give to others who are less fortunate. Thanks to Aida and Bill for their continued support and encouragement. They have dedicated their life to medicine and have always emphasized that knowledge is the key toward taking care of our patients and learning about oneself. We are endlessly appreciative of our many professors and teachers who have been unique role models in our lives. Finally, we thank our students and readers of this book and wish them a lifetime of happiness and education.

Preface

The medical literature is enormous. It is filled with information that is growing as you are reading this sentence. Having entered the realm of medical school and residency, it is very difficult to gather and master the information that is most critical. We have attempted to create a more concise and focused text that addresses the heart of the major issues that are not only commonly tested on Neurology and Neurosurgery Board exams but also encountered as one is taking care of sick patients. We have put much thought and effort into providing a text that can be used by pre-medical, medical students, nursing students, residents, as well as attending physicians and those in all scientific fields. Our goal is to provide an avenue of knowledge that can be used to heal that which is most important to us: Our Patients! Enjoy!

-Paul, Carolyn, and Sean

Note: All images are obtained from Authors (Kaloostian 2012).
Front cover pic: Large frontal tumor causing mass effect requiring surgical resection (Kaloostian 2012)

We are always looking to improve this book! Please email any questions, comments, or concerns regarding the book and improvements that can be made to this edition to the following email address: paulkaloostian@hotmail.com.

Table of Contents

Introduction

This is the first edition of the Pediatric Neurology Clinical Vignettes Textbook. This topic is truly fascinating. The cases described are real life cases experienced by the authors and each tells a different tale about pediatric neurological disorders. We present cases and detail questions that highlight unique aspects of the diagnosis and treatment of the case. Each of these stories has provided insight for the authors in their journey through medicine and we wanted to share these stories with you. We hope that as you read through this text you will come to learn more about diagnosing and treating a variety of pediatric neurological disorders. This material can be useful not only to residents in neurology and neurosurgery in preparation for the board examinations, but also residents in pediatrics and medicine, as well as medical students and nurses at all stages. Please enjoy!

-Paul, Carolyn, and Sean

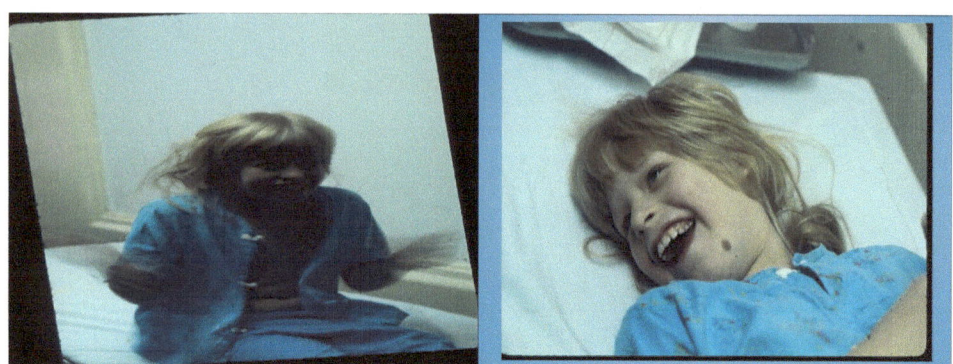

(Courtesy of Dr Mary Johnson, University of New Mexico Neurology Department)

What disorder does this young light-skinned blue-eyed girl have?

Phenylketonuria

What amino acid buildup causes this disorder?

Phenylalanine

What is treatment of choice?

Dietary restriction of Phe

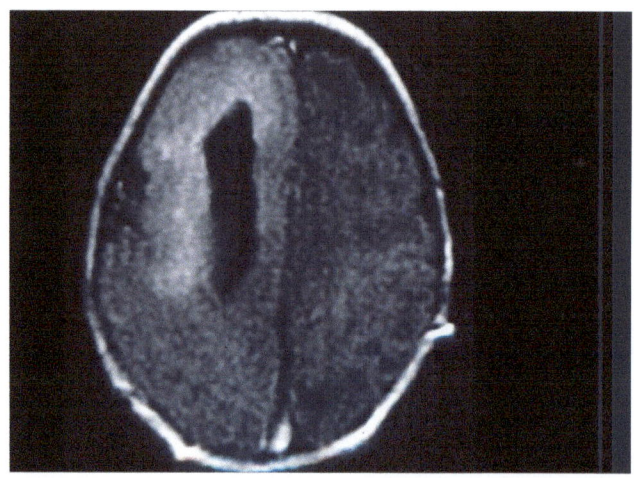

(Courtesy of Dr Mary Johnson, University of New Mexico Neurology Department)

What pathology is noted in the above MRI of this 5-year-old female with seizures?

Migrational disorder affecting right frontal lobe white and grey matter

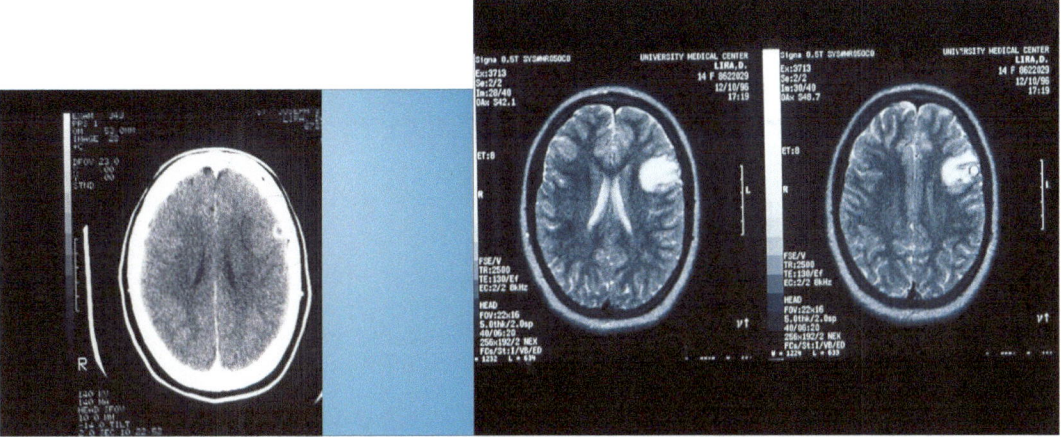

(Courtesy of Dr Mary Johnson, University of New Mexico Neurology Department)

What pathology is noted in the above CT and MRI in this 6-year-old patient from Mexico presenting with focal seizures?

Neurocysticercosis

What is treatment of choice?

Steroids + Albendazole
Repeat MRI in 3-5 weeks to measure interval change in size of cyst

What is treatment if repeat MRI shows enlarging cyst with worsening mass effect and continuing seizures?

OR for stealth guided cyst drainage

What food is the cause of this pathology?

Uncooked Pork

What organism is the cause of this pathology?

Tapeworm Teania solium

What aspect of this patient's history is commonly associated with this disorder?

Living in Mexico

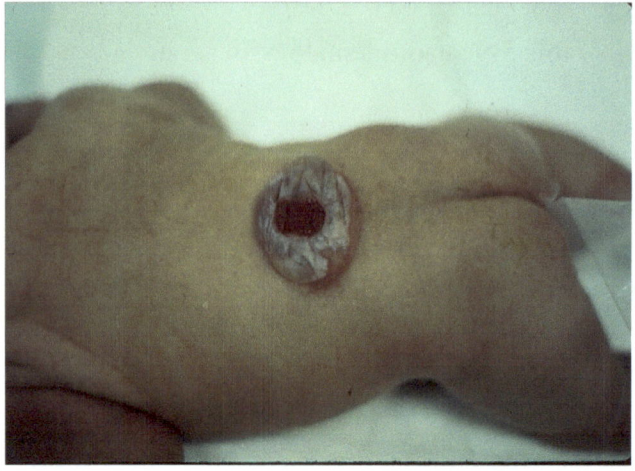

(Courtesy of Dr Mary Johnson, University of New Mexico Neurology Department)

What pathology is noted above?

Lumbosacral Myelomeningocele

What is treatment of choice for this patient?

OR for closure of myelomeningocele

What is best timing for surgery in patient's with this pathology?

As soon as possible to prevent meningitis

True or False. Neuromonitoring is often used during operative closure of this pathology.

False.

True or False. It is important to have a baseline neurological exam prior to surgery in these patients.

True

What other pathology is very frequently associated with the above condition?

Chiari malformation

What studies must be done in these patients after repair of myelomeningocele?

Ultrasound head-R/o hydrocephalus

Measurement of FOC and anterior fontanelle palpation

True or False. Most patients with the above pathology require a ventriculoperitoneal shunt.

True

True or False. Repairs of the above pathology can be done in utero.

True

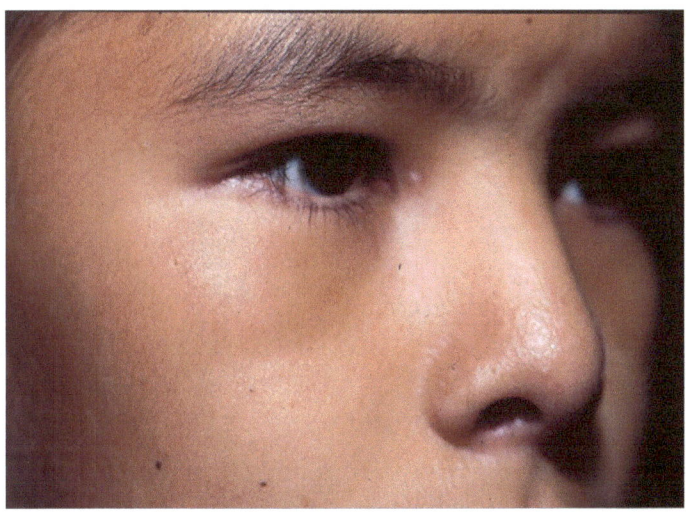

(Courtesy of Dr Mary Johnson, University of New Mexico Neurology Department)

What pathology is seen in the 5-year-old child above who presented with proximal muscle weakness?

Heliotrope rash under eyelids

What condition does this patient likely have?

Dermatomyositis

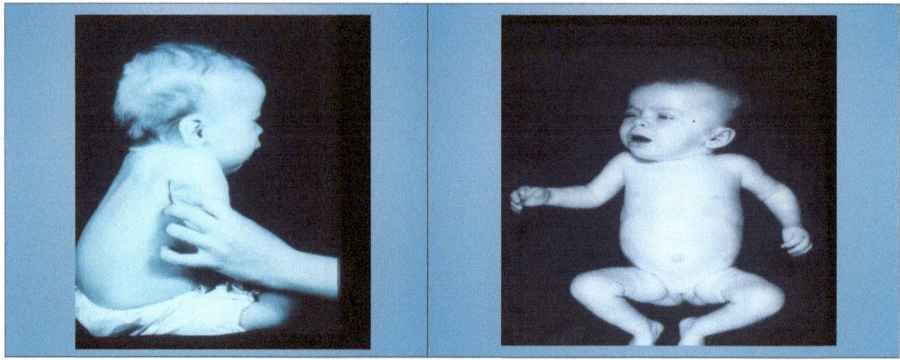

(Courtesy of Dr Mary Johnson, University of New Mexico Neurology Department)
6 months, 9 month

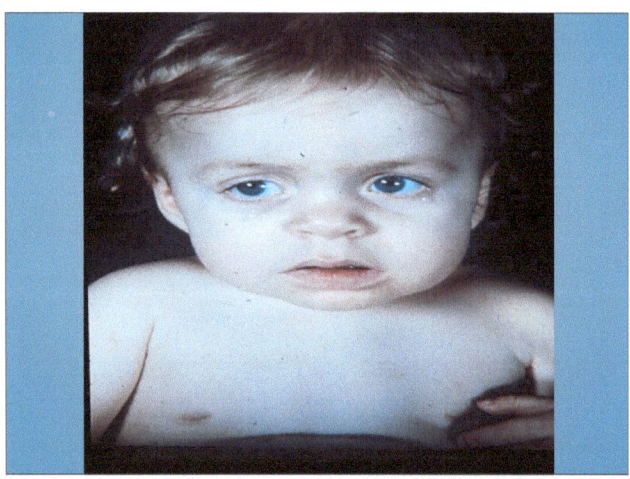

(Courtesy of Dr Mary Johnson, University of New Mexico Neurology Department)
13 months

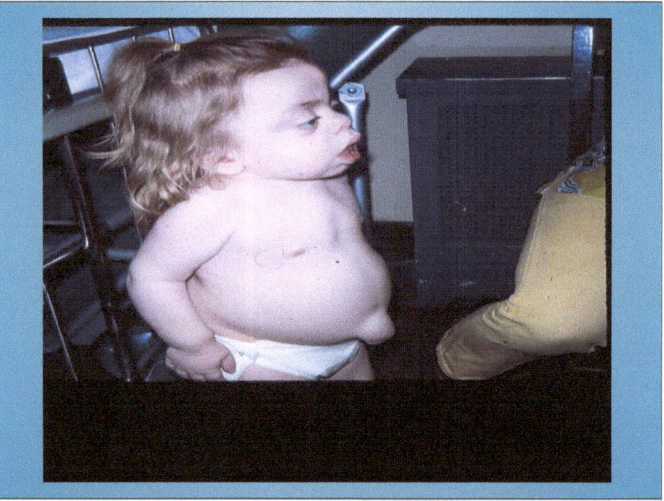

(Courtesy of Dr Mary Johnson, University of New Mexico Neurology Department)
3 years old

What pathology does this patient have?

Hurler's syndrome

What type of pathology is this disorder?

Mucopolysaccharidosis

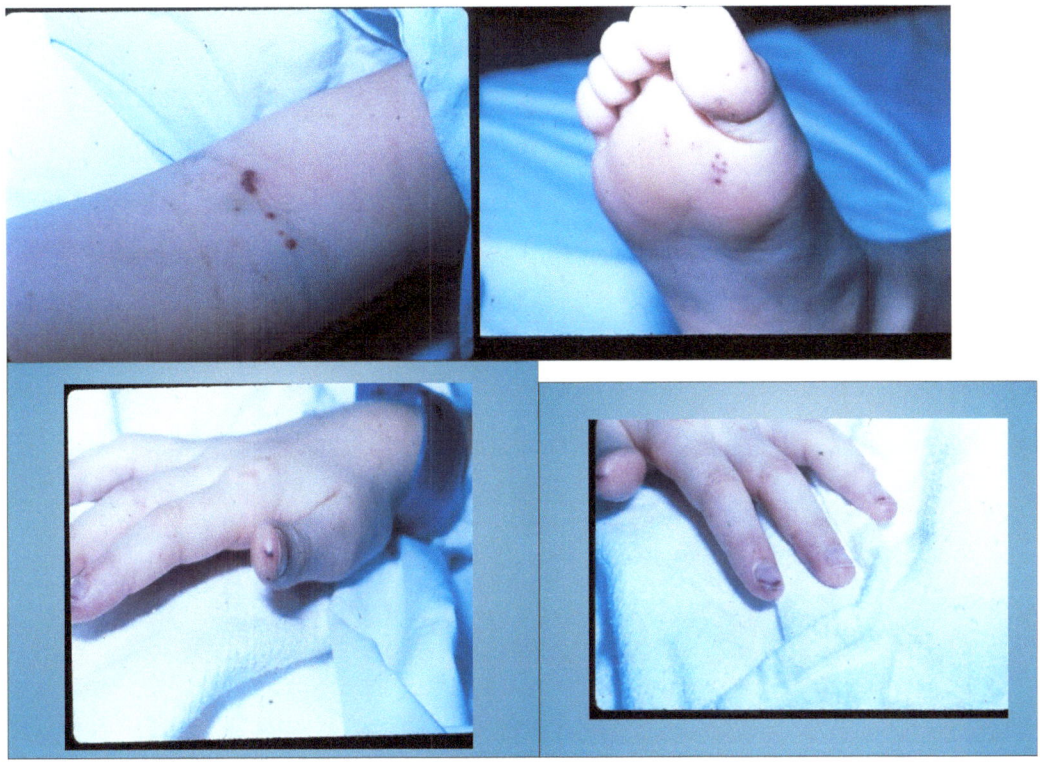

(Courtesy of Dr Mary Johnson, University of New Mexico Neurology Department)

4-year-old child presented with seizures and the above skin lesions. She has a recent history of endocarditis. What pathology does this young girl have?

Systemic lupus endocarditis (SLE)

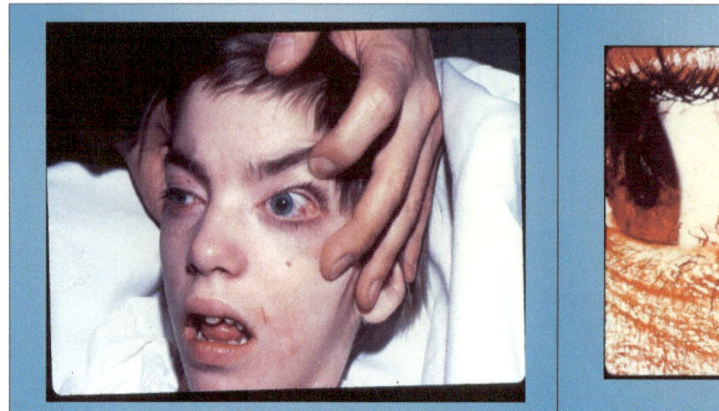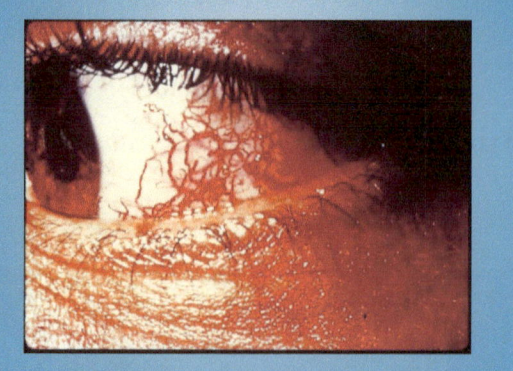

(Courtesy of Dr Mary Johnson, University of New Mexico Neurology Department)

8-year-old male presented with acute osnet of hypotonia, frequent respiratory infections, diminished growth, and severe ataxia of gait. He was initially diagnosed with cerebral palsy early on. On exam, he has the above findings with severe areflexia. What is the most likely diagnosis?

Ataxia telangiectasia

What genetic inheritance pattern does this disease follow?

Autosomal recessive

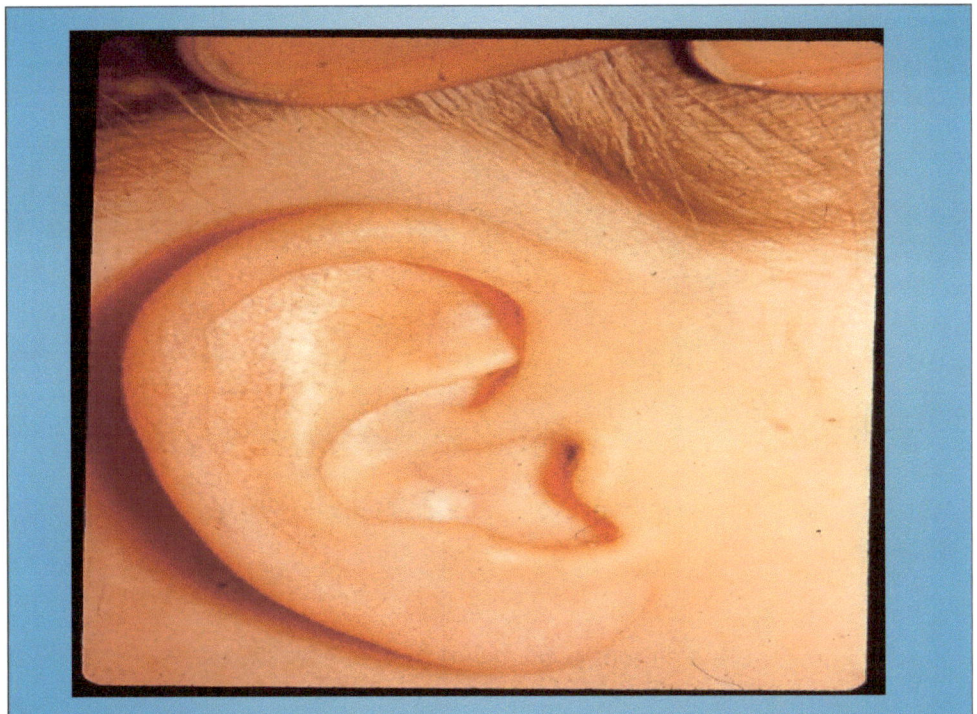

(Courtesy of Dr Mary Johnson, University of New Mexico Neurology Department)

2-year-old male is brought in by parents due to the above finding along both ears, as well as other places throughout his body. He has had multiple respiratory infections. His father also has this syndrome. What is the likely diagnosis?

Ataxia telangiectasia---telangiectasias of the ears

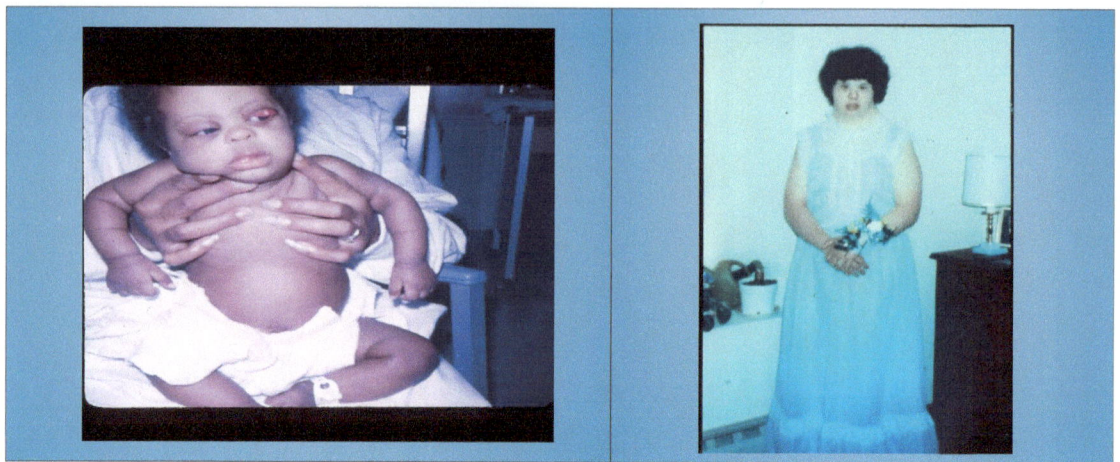

(Courtesy of Dr Mary Johnson, University of New Mexico Neurology Department)

What pathology is seen above in this patient at birth and then at age 17?

Trisomy 21 (Downs Syndrome)

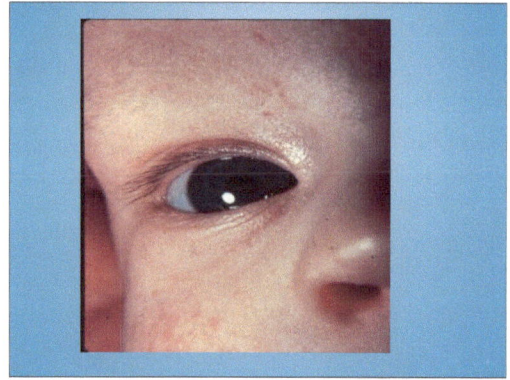

(Courtesy of Dr Mary Johnson, University of New Mexico Neurology Department)

What pathology is seen in this patient with Trisomy 21?

Brushfield spots under the eyes

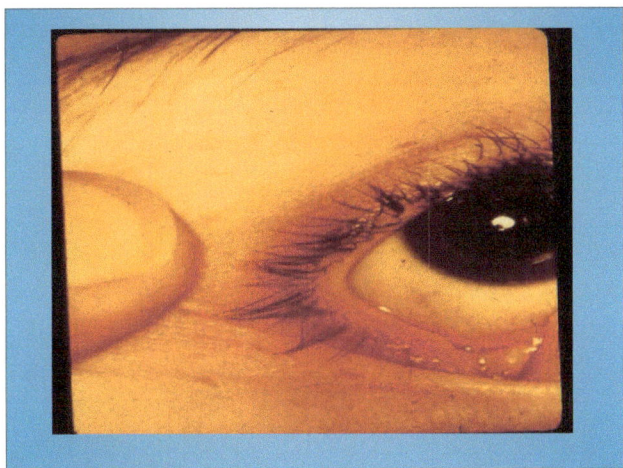

(Courtesy of Dr Mary Johnson, University of New Mexico Neurology Department)

What is the likely diagnosis in this 3-year-old male with seizures and an abnormality in the eyes seen best with slit-lamp examination?

Wilson disease

What is the term given to the pathology within the eye?

Kaiser-Fleischer Rings

What is the genetic inheritance pattern?

Autosomal recessive

What gene is abnormal in these patient's?

ATP7B gene

What element is abnormally accumulating in this pathology?

Copper

What organs systems are typically affected in this pathology?

Liver, kidneys, cardiac myopathy, Eyes, psychiatric disorder,

What chromosome is the ATP7B gene on?

13

What does this gene encode?

ATP ase that transports copper into bile to incorporate into ceruloplasmin for excretion

In this pathology, ceruloplasmin levels are typically_____?

Low

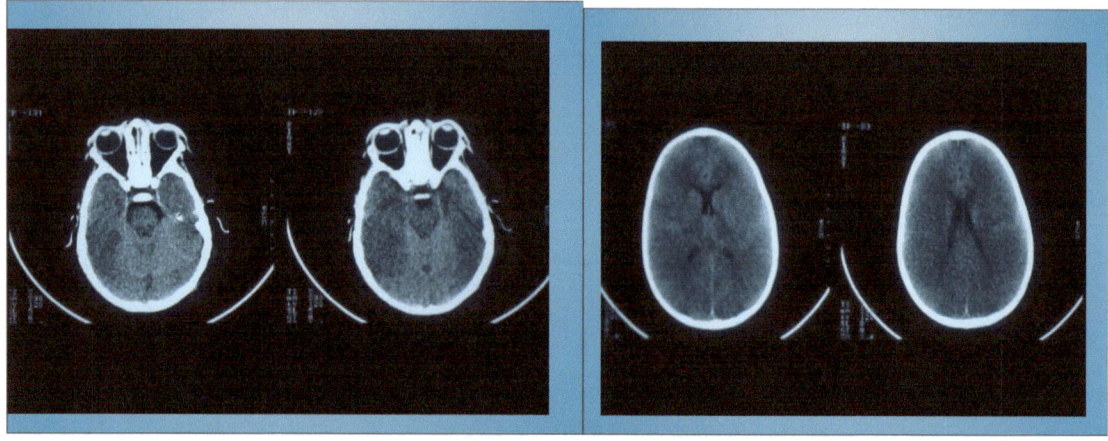

(Courtesy of Dr Mary Johnson, University of New Mexico Neurology Department)

3-year-old female is involved in an MVC that has killed her father and mother. She is hypoxic on seen and required coding to rescusitate her. Her neuro exam demonstrates no brainstem reflexes. She has triple flexion in her lower extremities. The CT scans are shown above. What is the likely diagnosis?

Anoxic injury to the brain—diffuse loss of grey-white differentiation throughout

What phenomenon is seen in the above scans that is often seen in severe anoxic injury?

Cerebellar reversal sign—cerebellum more hyperdense that the rest of the brain (usually the opposite in normal patients)

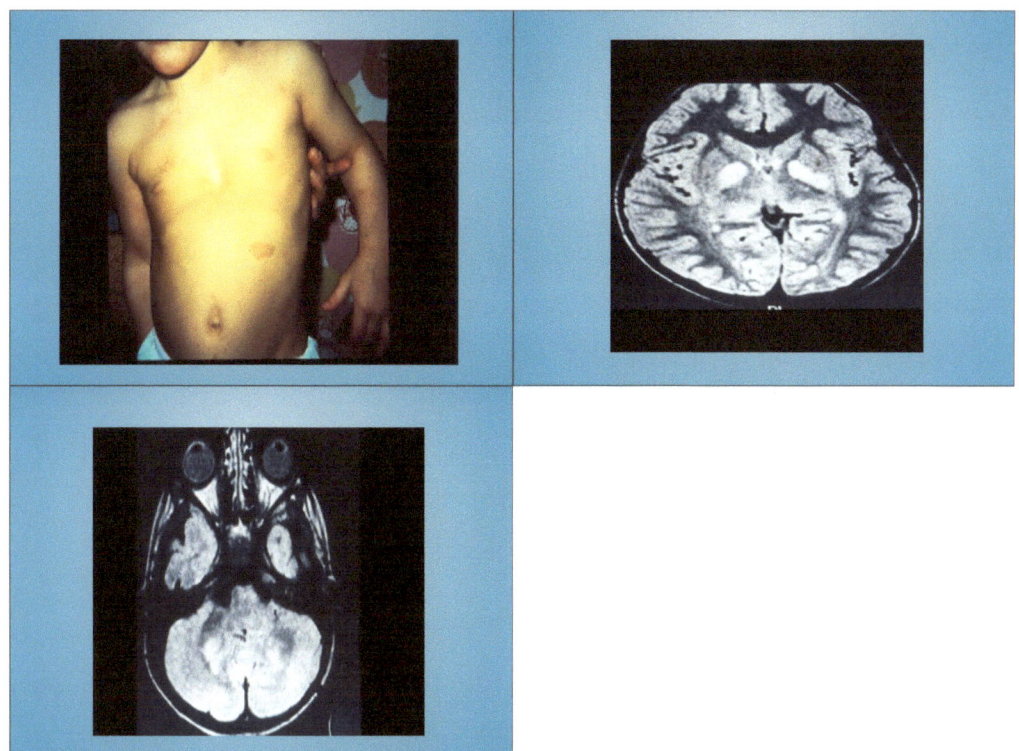

(Courtesy of Dr Mary Johnson, University of New Mexico Neurology Department)

What is the likely diagnosis in the above patient with the skin findings and MRI brain, as well as a mother with the same pathology?

Neurofibromatosis I

What is seen on the above MRI scans?

UBO—unidentified bright objects

Which of the following is seen in NF-I?

Optic gliomas
Lisch nodules
Neurofibromas (>2) or one plexiform neurofibroma
Sphenoid dysplasia
Pseudoarthrosis of long bone cortex
Caf au lait >6

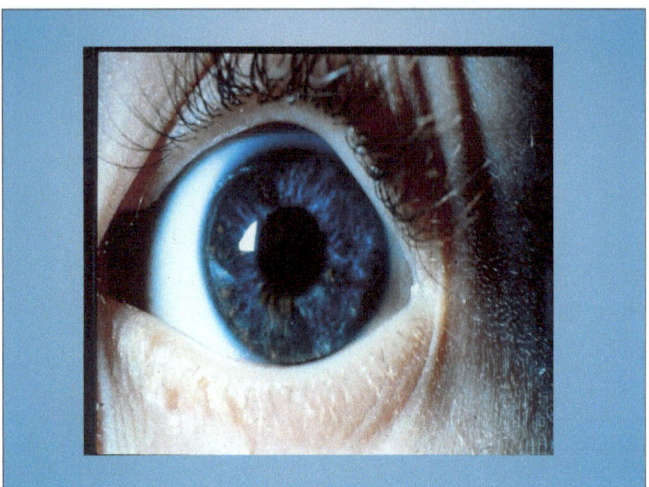

(Courtesy of Dr Mary Johnson, University of New Mexico Neurology Department)

What is the diagnosis in this patient with above physical exam finding and a family history of this lesion?

NF-1

What are these called?

Lisch nodules

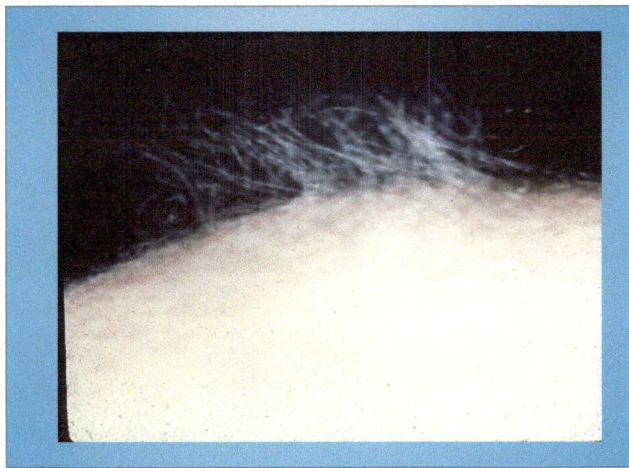

(Courtesy of Dr Mary Johnson, University of New Mexico Neurology Department)

3-year-old patient with very thin brittle hair that easily comes out. Upon closer view, the hair is actually in a twisted pattern. What is the name of this pathology?

Menke's Disease (Kinky Hair Disease)

What is the inheritance pattern in this disease?

X lined recessive

What element is abnormally low?

Copper

What other findings are associated with these patient's?

Sparse and coarse hair, hypotonia, growth failure, fair skin, seizures

What is the term given to the structure of hair upon microscopic examination descrbied by Datta et.al 2008 ?

Pili Torti

What gene is abnormal in these patient's?

ATP7A gene

What is the syndrome associated with Menkes disease where the occipital bone has significant calcium deposits with coarse brittle hair and loose joints?

Occipital horn syndrome

What is treatment?

IV or SubQ injections of Copper

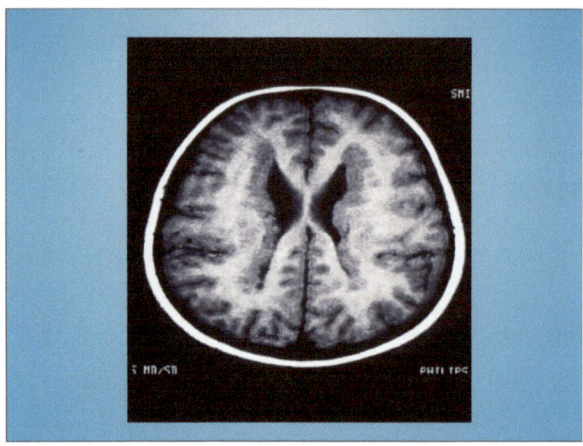

(Courtesy of Dr Mary Johnson, University of New Mexico Neurology Department)

What is the term for the above pathology in this 5-year-old male with seizures and mental retardation?

Heterotopia

What type of abnormality causes this pathology?

Neuronal migration defects

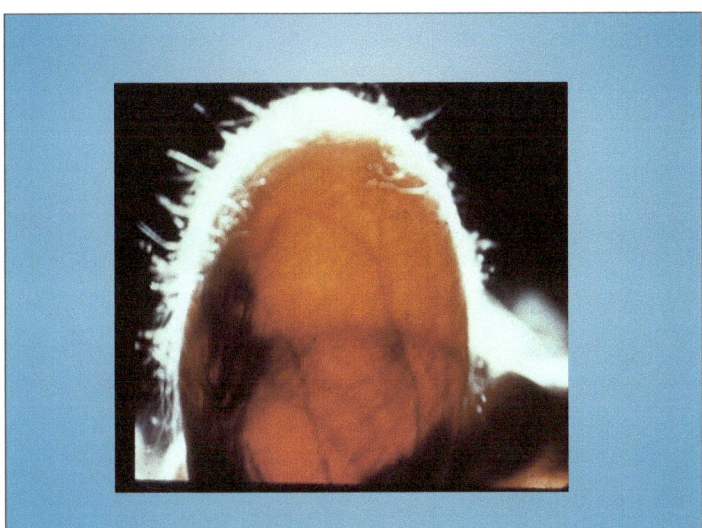

(Courtesy of Dr Mary Johnson, University of New Mexico Neurology Department)

What physical exam is being done in this newborn patient?

Transillumination of the skull

What is the likely diagnosis?

Hydranencephaly

Place Chun gun pic here and the other….

What are these devices used for?

Transillumination of the skull

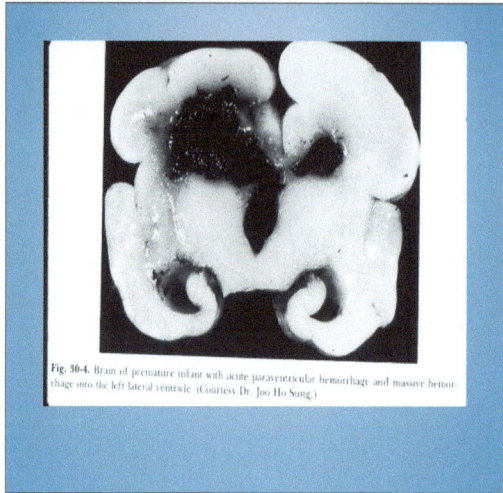

Fig. 50-4. Brain of premature infant with acute paraventricular hemorrhage and massive hemorrhage into the left lateral ventricle. (Courtesy Dr. Joo Ho Sung.)

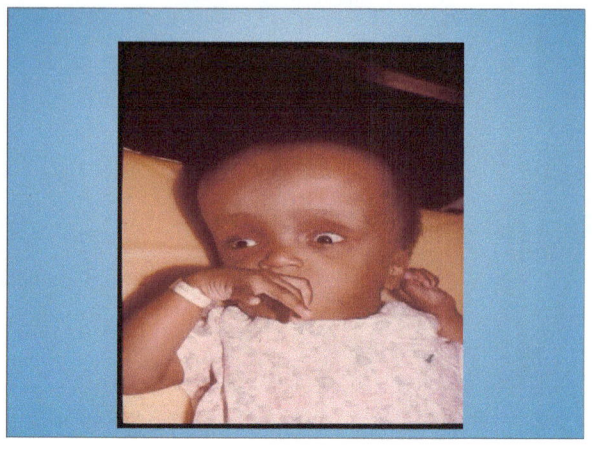

(Courtesy of Dr Mary Johnson, University of New Mexico Neurology Department)

What is the diagnosis in this patient with the physical findings shown above, tense anterior fontanelle and increasing FOC?

Hydrocephalus

What grade of hemorrhage is shown above on the post-mortem autopsy?

Grade 4—both IPH and IVH

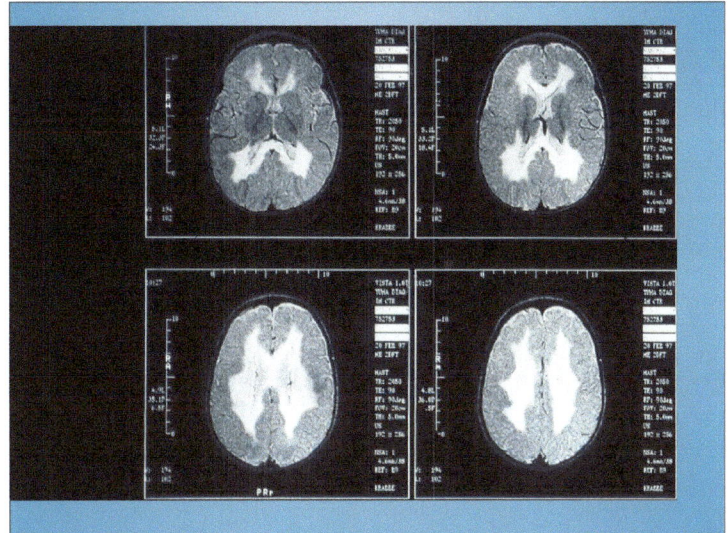

(Courtesy of Dr Mary Johnson, University of New Mexico Neurology Department)

What is the diagnosis in this 1-year-old male with fevers, mental retardation, hypotonia, seizures, and motor delay?

Krabbes Disease—both anterior and post white matter are affected (adrenoleukodystropy—only post white matter affected)

What is the genetic inheritance?

Autosomal recessive

What country has the highest incidence of this pathology?

Isreal

What gene is affected?

GALC gene

What chromosome is this on?

14

What enzyme is deficient?

Galactocerebrosidase

What aspect of the CNS is affected in this disease?

Myelin

What characteristic finding is seen on luxol fast staining?

Minimal myelin and clustering of globoid multinucleate cells

NEJM recently reported what finding in relation to treatment of these patients?

If treat with core blood transplants before symptoms, extend survival

What other treatment may be done?

Bone marrow transplant

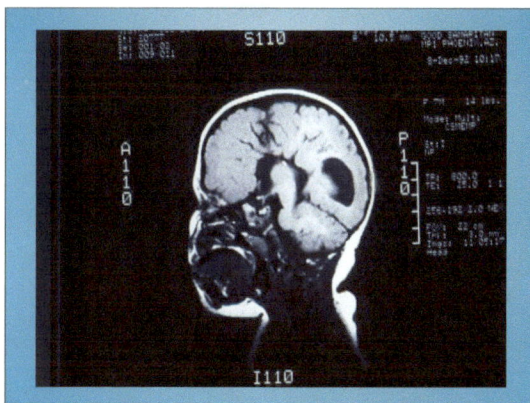 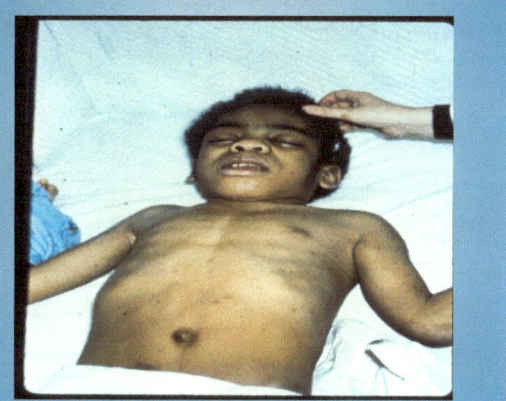

(Courtesy of Dr Mary Johnson, University of New Mexico Neurology Department)

What pathology is seen in this patient with pica, mental and motor delays, and progressive dementia?

San Filippo syndrome

What is genetic inheritance?

Autosomal recessive

What organelle abnormality is the cause of this disorder?

Lysosomal storage disease

What glycosaminoglycan is abnormally accumulated in this disorder?

Heparin sulfate

There are 4 types of San Filippo syndromes. Which chromosome have abnormalities associated with them?

8,12, 17

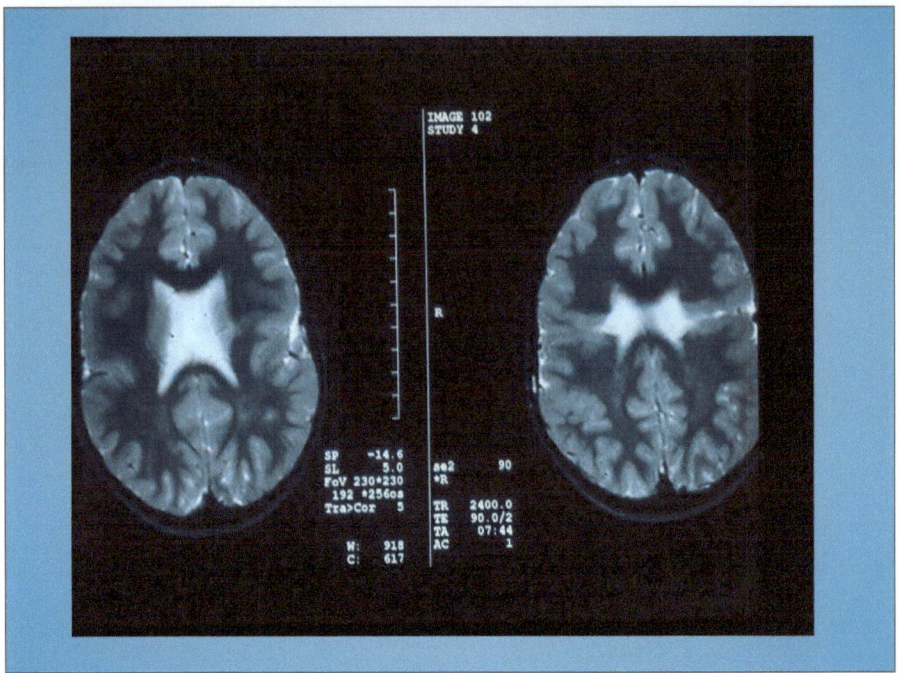

(Courtesy of Dr Mary Johnson, University of New Mexico Neurology Department)

What is the term for the pathology above?

Schizencephaly

The tracts in this disease are often lined by what?

Grey matter

This disorder is due to what problem in utero?

Abnormal neuronal migration

What disorders in utero can predispose to this pathology?

EBV, CMV, medications taken by mother

What is the counterpart of this disorder that is characteristically seen post-trauma with a large fluid collection lined by white matter?

Porencephaly

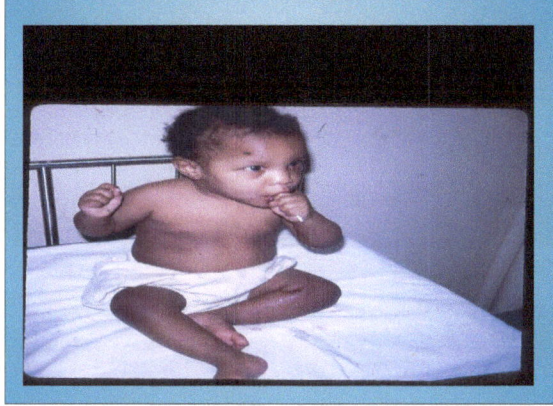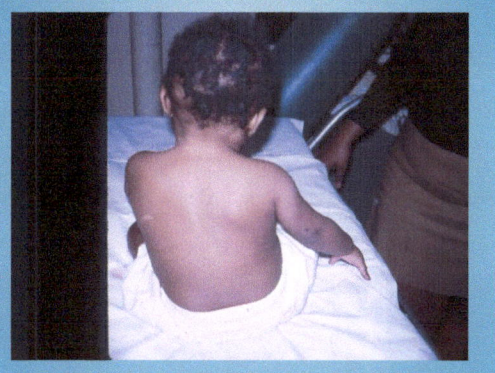

(Courtesy of Dr Mary Johnson, University of New Mexico Neurology Department)

What pathology is seen in this child with the above depigmented spots and seizures, with a family history of this pathology?

Tuberous Sclerosis

How many types of this pathology are noted?

2

What genes and associated proteins are encoded in the specific types?

TSC1—Hamartin
TSC 2—Tuberin

What chromosomes are associated with TSC1?

9

TSC 2?

16

What lesions are seen in this pathology?

Cortical tubers, subependymal nodules, angiomyolipomas, lung cysts, cardiac rhabdoymyomas, facial angiofibromas, subungual fibromas, ash-leaf spots (seen under woods lamp), shagreen thich patches on back

What cranial tumors are seen in this disease?

Subependymal giant cell astrocytomas (SEGA)

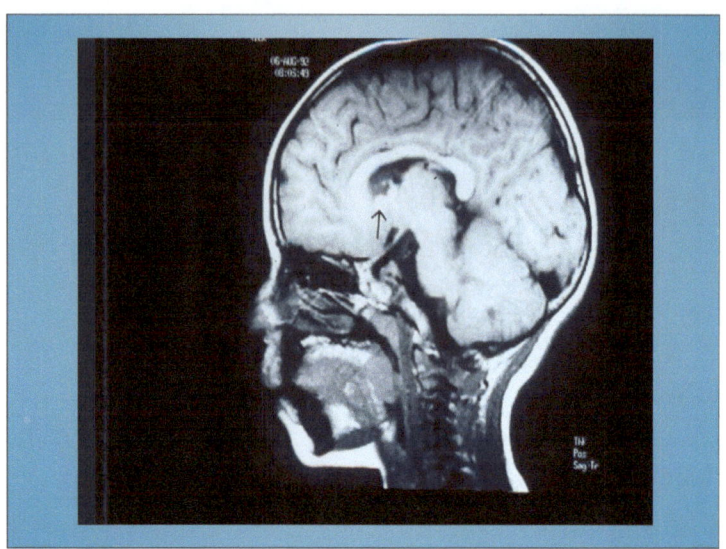

(Courtesy of Dr Mary Johnson, University of New Mexico Neurology Department)

What pathology is noted above in this child with seizures and hypomelanotic macules?

Subependymal tubers or hamartomas

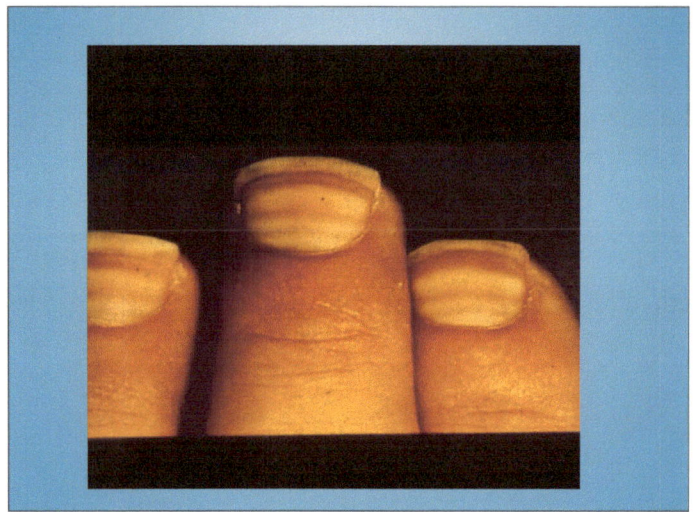

(Courtesy of Dr Mary Johnson, University of New Mexico Neurology Department)

What pathology is seen in this young child with acute onset of nausea and vomiting and seizures with the above finding?

Mees lines

What elemental overdose is seen in patient's with this pathology?

Arsenic poisoning

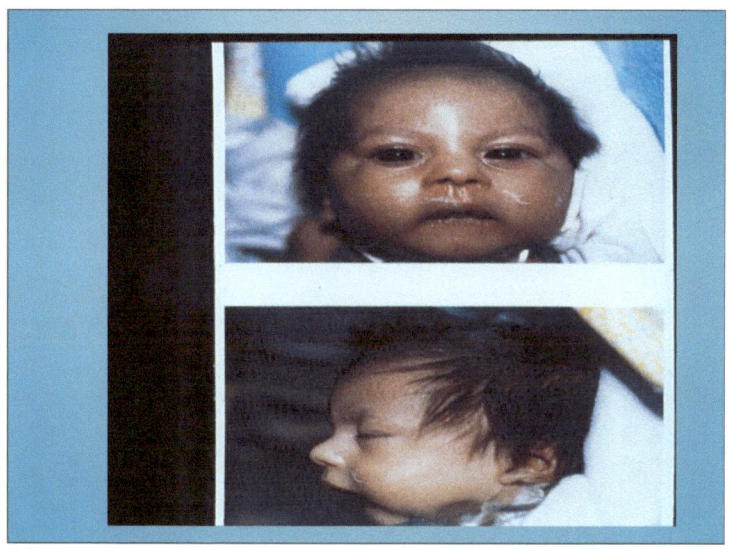

(Courtesy of Dr Mary Johnson, University of New Mexico Neurology Department)

What syndrome is seen in this young child?

Miller-Dieker Syndrome

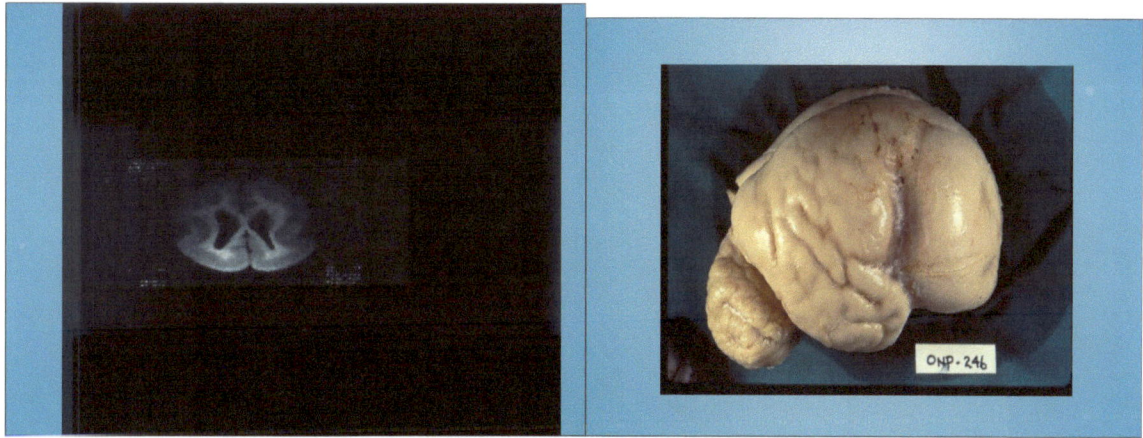

(Courtesy of Dr Mary Johnson, University of New Mexico Neurology Department)

What pathology is seen in this young child with mental retardation and seizures?

Lissencephaly

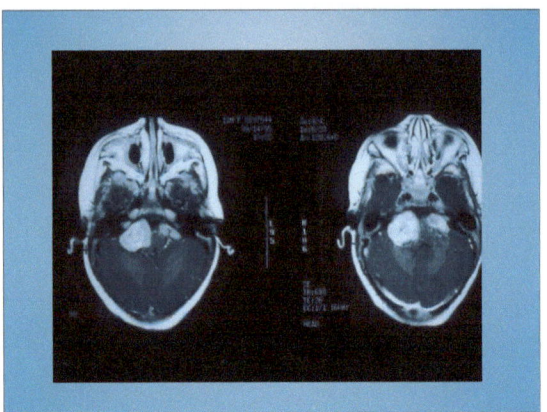

(Courtesy of Dr Mary Johnson, University of New Mexico Neurology Department)

What pathology is associated with the above MRI brain in this young child with hearing loss bilaterally and a family history of this disorder?

NF-2

What is the genetic inheritance pattern?

Aut dominant

What chromosome is defective in this patient?

22

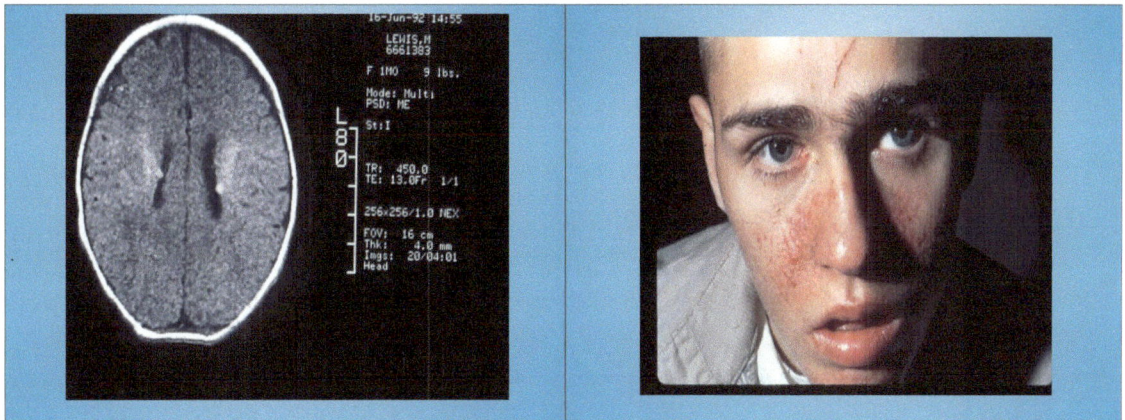

(Courtesy of Dr Mary Johnson, University of New Mexico Neurology Department)

What disorder is associated with the above findings?

Tuberous sclerosis

(Courtesy of Dr Mary Johnson, University of New Mexico Neurology Department)

What pathology is associated with the above phenotype in a family with a history of alcohol abuse?

Fetal alcohol syndrome

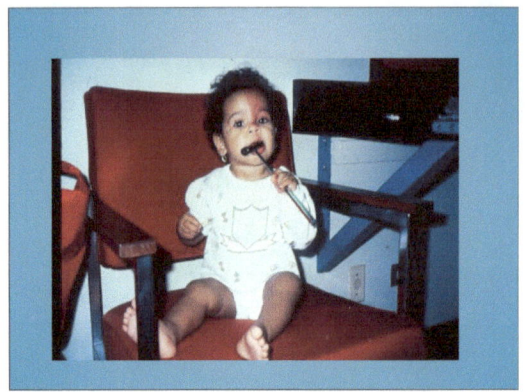

(Courtesy of Dr Mary Johnson, University of New Mexico Neurology Department)

What pathology does this child have?

Sturge Weber syndrome

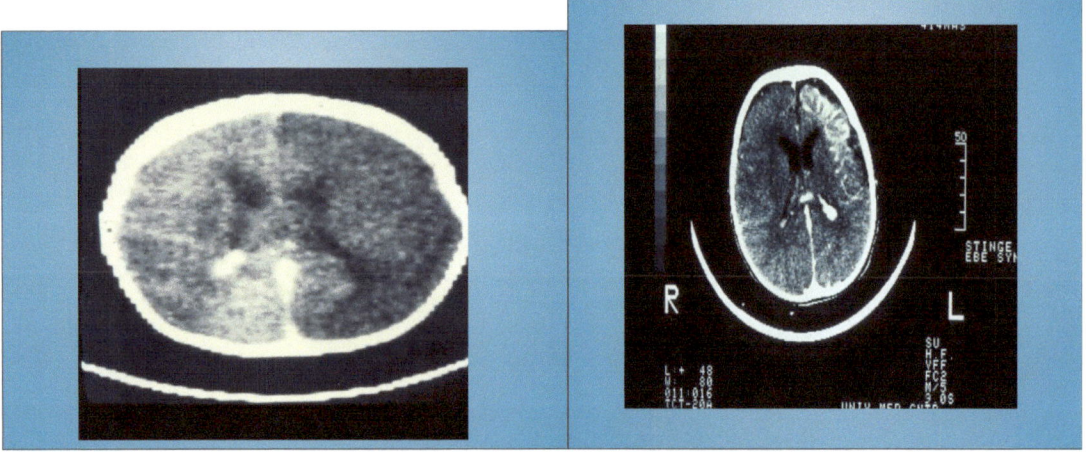

(Courtesy of Dr Mary Johnson, University of New Mexico Neurology Department)

What pathology is associated with the above scans in a young child with a rash, right seizures and hemiparesis?

Sturge weber syndrome

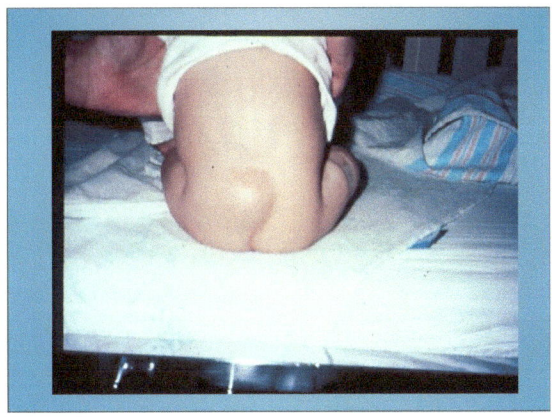

(Courtesy of Dr Mary Johnson, University of New Mexico Neurology Department)

What pathology is seen in the above patient?

Lipomyelomeningocele

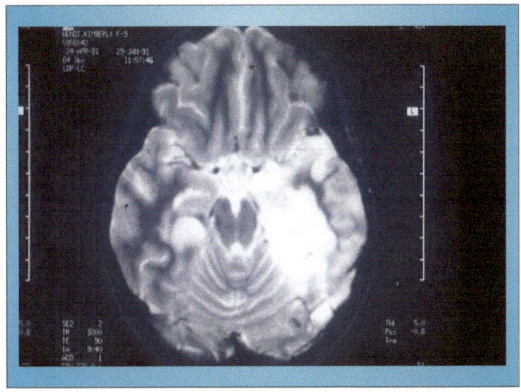

(Courtesy of Dr Mary Johnson, University of New Mexico Neurology Department)

What is the likely pathology in this newborn with seizures, hypotonia, and fevers?

Herpes simplex infection with hemorrhage in medial temporal lobe left

What must be started stat?

Acyclovir IV

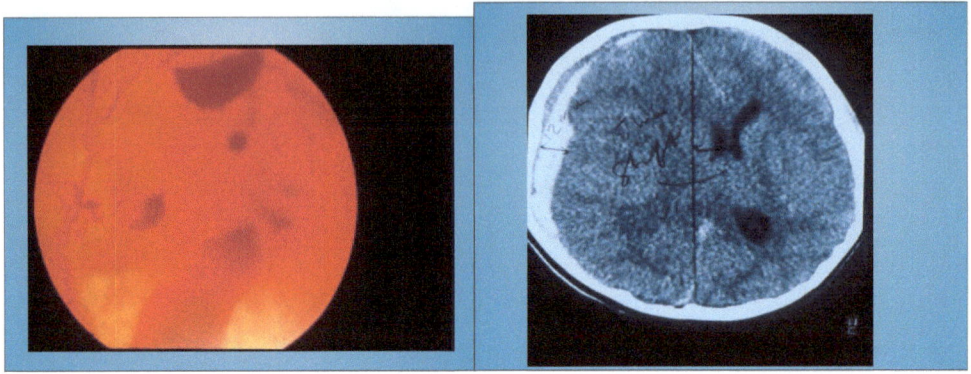

(Courtesy of Dr Mary Johnson, University of New Mexico Neurology Department)

What is the likely pathology in this 6-month-old male?

Non-accidental trauma causing retinal hemorrhages and SDH

What is treatment for this child?

OR for right craniotomy for SDH evac

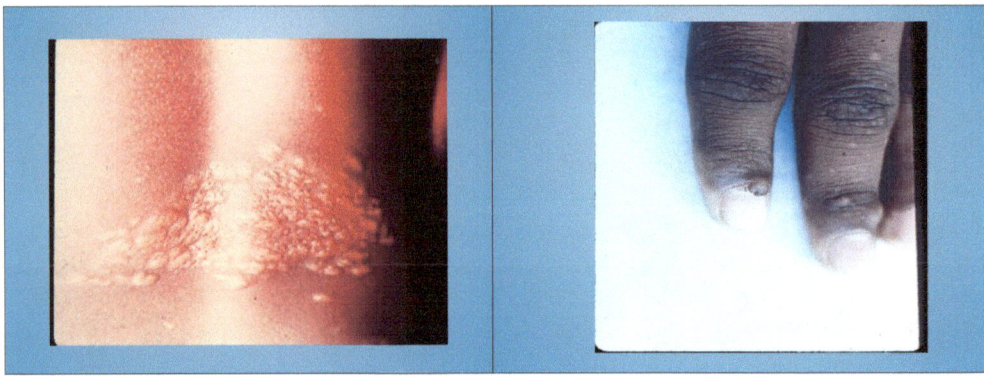

(Courtesy of Dr Mary Johnson, University of New Mexico Neurology Department)

What is the likely diagnosis in this child with above findings?

Tuberous sclerosis

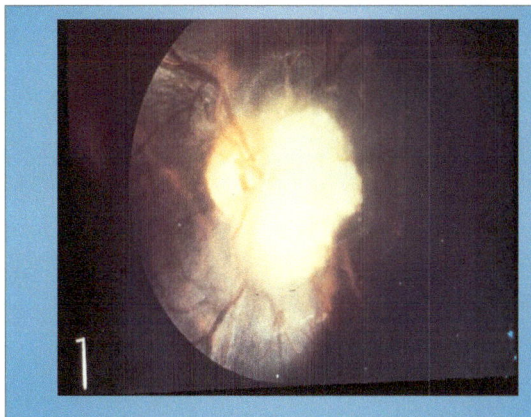 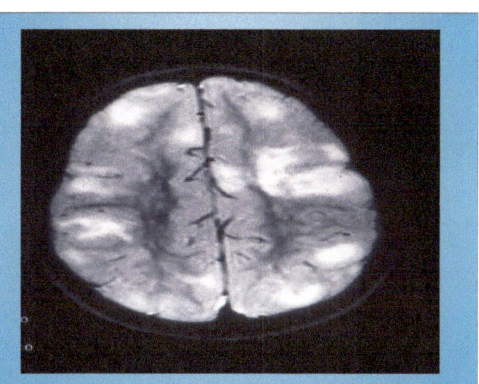

(Courtesy of Dr Mary Johnson, University of New Mexico Neurology Department)

What pathology is seen in the above images in this child with seizures?

Tuberous sclerosis

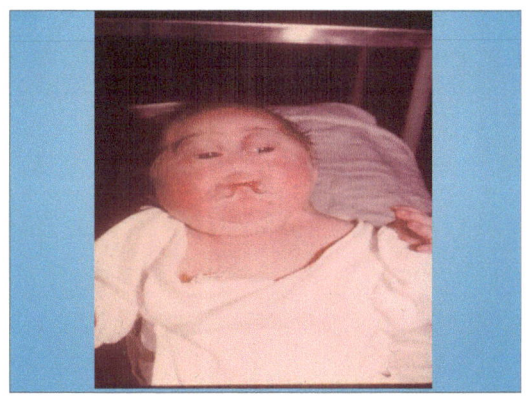

(Courtesy of Dr Mary Johnson, University of New Mexico Neurology Department)

What pathology is seen in the above patient?

Holoprosencephaly

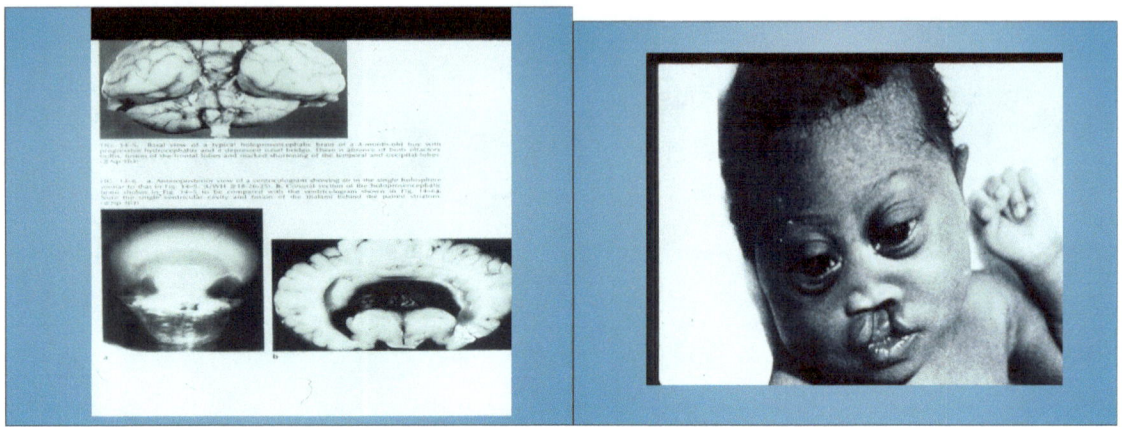

(Courtesy of Dr Mary Johnson, University of New Mexico Neurology Department)

What pathology is associated with the above findings in this young child?

Holoprosencephaly

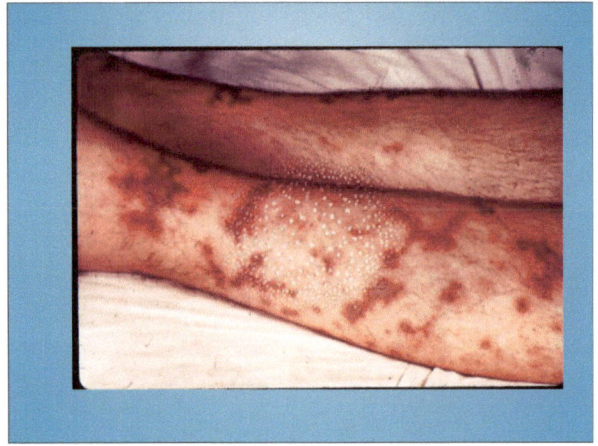

(Courtesy of Dr Mary Johnson, University of New Mexico Neurology Department)

What pathology is associated with this 18-year-old female boarding school student with acute onset of fevers, chills, nausea, and renal insufficiency?

Neisseria meningococcemia

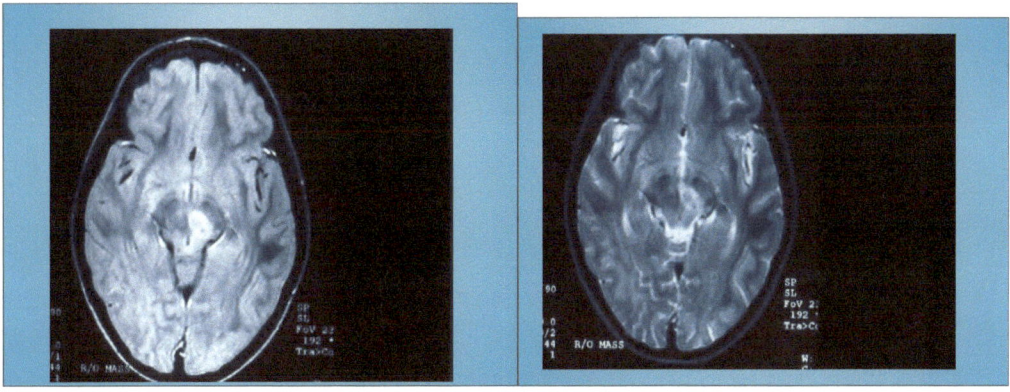

(Courtesy of Dr Mary Johnson, University of New Mexico Neurology Department)

What is the diagnosis in this newborn with seizures and minimal neurological status born to a prostitute mother without prenatal care?

HIV encephalopathy of the newborn

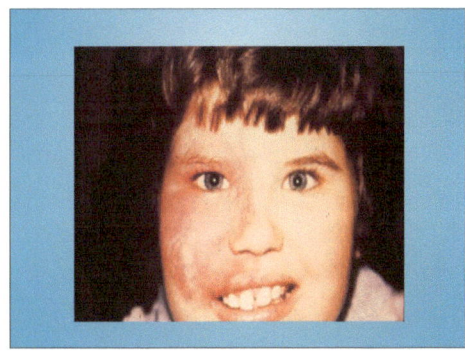

(Courtesy of Dr Mary Johnson, University of New Mexico Neurology Department)

What is the diagnosis in this child with hemiparesis and seizures?

Sturge-Weber syndrome

(Courtesy of Dr Mary Johnson, University of New Mexico Neurology Department)

What is the diagnosis in this patient with CT showing a large cisterna magna?

Dandy walker syndrome

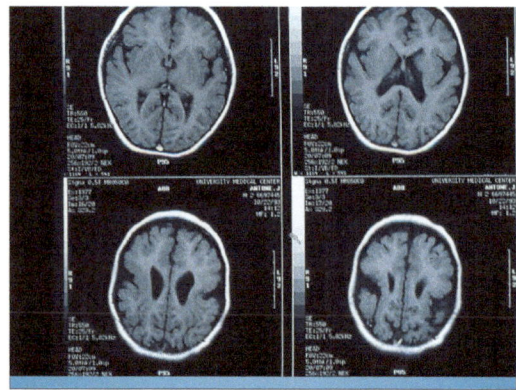

(Courtesy of Dr Mary Johnson, University of New Mexico Neurology Department)

What is the diagnosis in this young child with seizures?

Schizencephaly

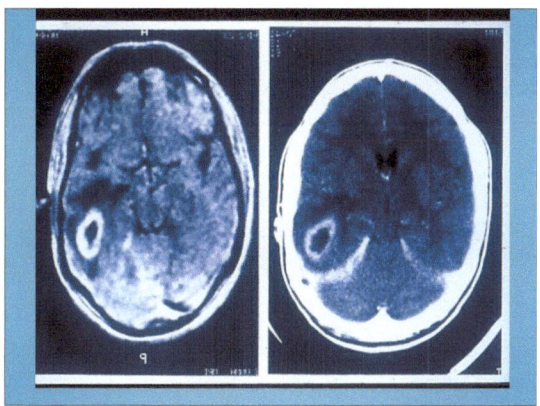

(Courtesy of Dr Mary Johnson, University of New Mexico Neurology Department)

What is the diagnosis in this young child with fevers, chills, and an elevated wbc count with new onset altered mental status?

Intracranial right temporal lobe abscess

What is treatment?

Stealth guided drainage followed by Abx

What childhood disease is associated with multiple vascular anomalies and cranial infections?

Ataxia telangiectasia

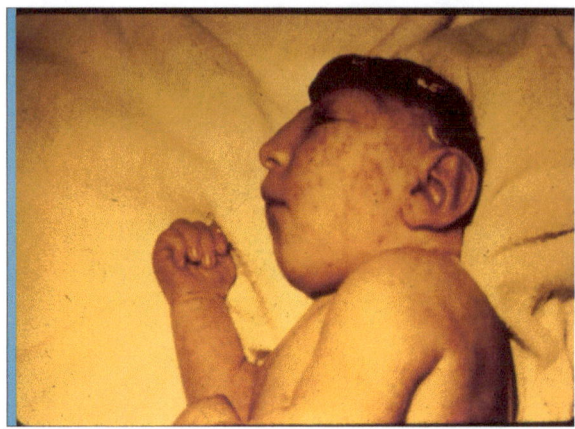

(Courtesy of Dr Mary Johnson, University of New Mexico Neurology Department)

What pathology is seen in the above patient with minimal neurological exam?

Ancencephaly

What pre-natal vitamin deficiency is associated with this?

Folic acid

What lab value is often elevated in pregnant woman with this pathology?

AFP

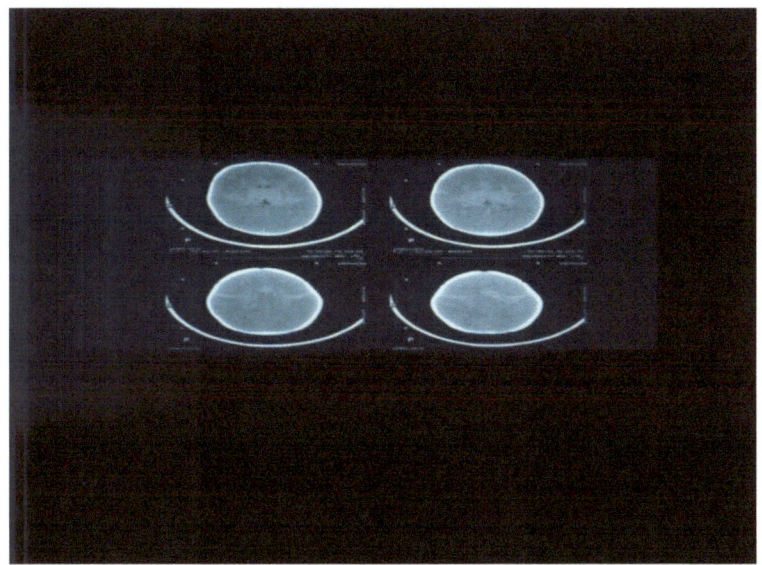

(Courtesy of Dr Mary Johnson, University of New Mexico Neurology Department)

What is seen in this child with status epilepticus for 10 minutes?

Diffuse anoxic injury

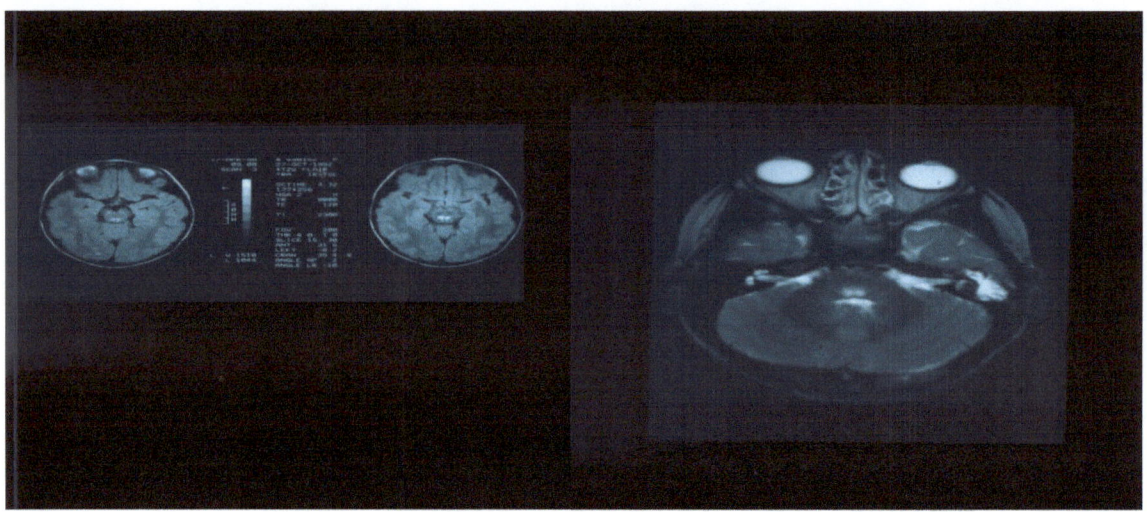

(Courtesy of Dr Mary Johnson, University of New Mexico Neurology Department)

What pathology is seen in this patient with a maternally inherited defect causing hypotonia and seizures with MRI demonstrating brainstem lesions?

Leigh's disease

(Courtesy of Dr Mary Johnson, University of New Mexico Neurology Department)

What disorder is seen in this patient with constant hand-writhing motions, with loss of useful hand use, loss of coloring ability, and difficulty writing?

Rette Syndrome

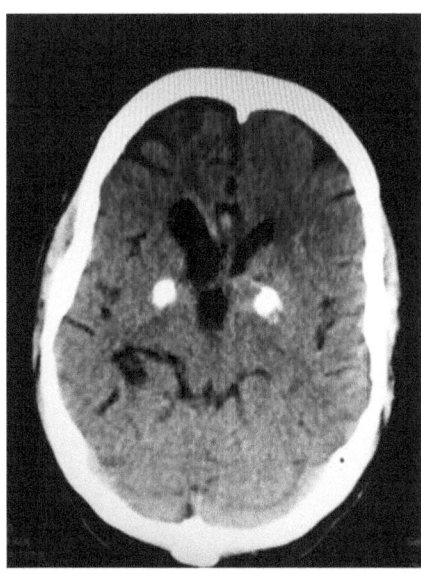

34-year-old male presents with seizures. He is neurologically intact on examination. CT Head is shown above. What disorder is this typically seen in?

Fahr disease

What studies should be done next?

EEG, MRI

What consultation should be obtained?

Neurology

Where are the above calcifications exactly?

Bilateral globus pallidus interna

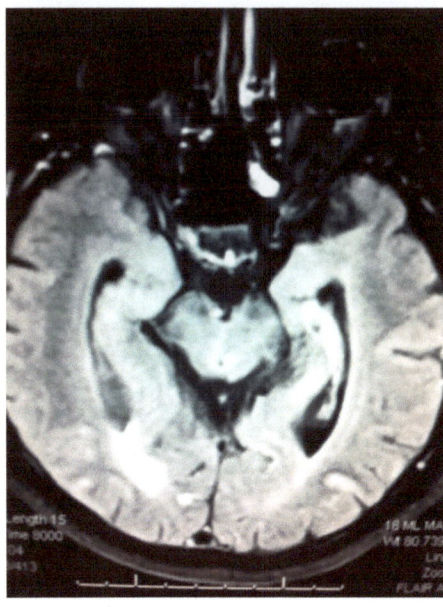

37-year-old female presents with fevers, chills, and headaches. She is obtunded. MRI brain is shown above. What is the differential diagnosis for the above findings?

Septic emboli, Demyelinating disorder, Postinfectious encephalopathy

What are the next steps in management?

LP, Pan cultures, empiric antibiotics

What disorder should be watched for in a young patient presenting with altered mental status and seizures with a temporal lobe hemorrhage or signal change?

Herpes encephalitis

What medication should be started immediately if the above is suspected?

Acyclovir IV

How would one diagnose this disorder?

LP + for HSV PCR

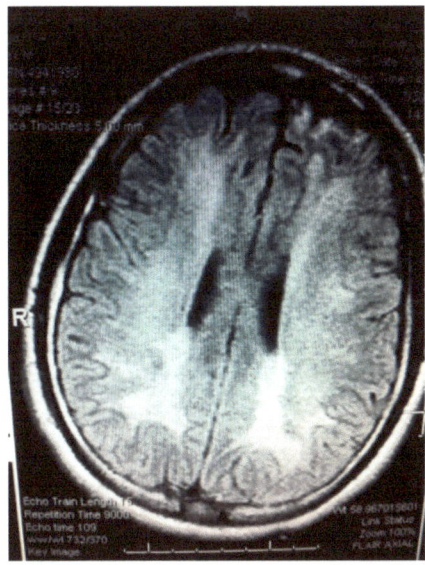

11-year-old female ingested large quantity of her father's morphine sulfate. She is obtunded. She had no episodes of apnea or cardiac arrest per report. MRI is shown above. What is the likely diagnosis?

Leukoencephalopathy

What is treatment?

Supportive

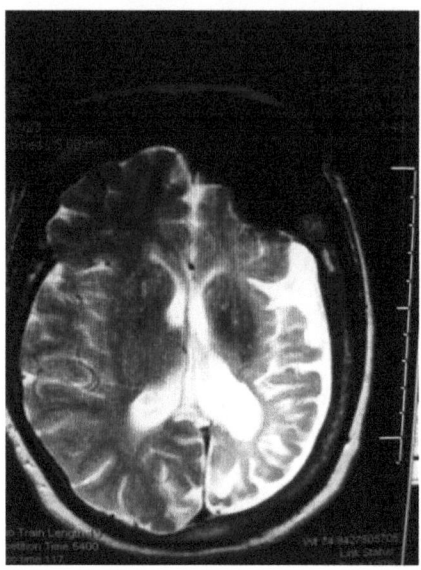

What is the diagnosis in this patient with large left frontal sinus, left hemiatrophy of brain, and thicked marrow of skull on left side?

Dyke-Davidoff-Masson syndrome

SUMMARY:

We hope that after reading this textbook one would be able to have a much more clear idea of the diversity of pediatric neurological disorders as well as gain a better understanding of some of the more rare disorders in this field. This text can be used to review a variety of neurological disorders that may show up on board examinations or in our practice. The accompanying textbook "Cracking Pediatric Neurosurgery Vignettes" should be utilized in conjunction with this textbook for a more complete pediatric neurology and neurosurgery experience. Please enjoy!

The Authors

(Back cover pic: cerebellar hemangioblastomas as seen in a patient with von-hippel-lindau syndrome- Kaloostian 2012)

Hippocratic Oath

I swear by all that I hold most sacred
That I will keep this enduring oath:

To the best of my ability and judgement I will
Practice the Art only for the benefit of my patients.

Whatever houses I may visit, I will enter only
To help the sick or to prevent illness,
Never to inflict harm, injustice, or suffering.

I will lead my life and practice the Art
Conscientiously and with honor.

Whatever I may see or hear in the practice of the Art
Or even outside of it that should not be spread abroad
I will keep in solemn confidence.

I will be just and generous to those who taught me
The Art, to my colleagues in the Art, and to
Those who desire to learn it.

May happiness and the physician's good repute be
Granted me while I keep this sacred Oath inviolate.

www.ingramcontent.com/pod-product-compliance
Lightning Source LLC
Chambersburg PA
CBHW051059180526

45172CB00002B/701